Stress in Crop Plants

Dr. Mubarak Idris

Kelsiehanson Verlag

DEDICATION

To My Wonderful Readers

CONTENTS

ACKNOWLEDGMENTS

Thank you to everyone who strives to grow and help others grow

1 ABSTRACT

Abiotic stress is the main source of danger for many crop plants. Salinity and droughts are the most common categories of this stress all over the world. In this study, Transformation of *EsWRKY11* in *Oryza sativa* (Giza 177) enhanced salt tolerance at the seedling stage. As *Oryza satvia*plant is considered one of the most important crops, it ranks the most second important cereal in the world. *Oryza satvia* production represents 30% of the world's cereal production today. After many searches, it was found that the *WRKY* transcription factors play an important role in the regulation of transcriptional reprogramming associated with plant abiotic stress responses. In this study, the *WRKY* transcription factor *EsWRKY11*, containing the plant-specific *WRKY* zinc finger DNA-binding motif, was isolated from *Eruca sativa*. The source of the *WRKY11* gene is members of family *Brassicaceae* such as *Arabidopsis thaliana* and *Eruca sativa*. Nanoparticles are a new trend technique applied for gene transformation. For example, chitosan nanoparticles are biocompatible, biodegradable, and give high transformation efficiency. The transformed *EsWRKY11* gene was detected in rice transgenic lines. *EsWRKY11* was upregulated in response to a variety of abiotic stresses, including salinity, alkalinity, cold, abscisic acid, and drought. Micropropagation tissue culture technique was used for the multiplication of both transgenic and non-transgenic rice lines. Transgenic *Oryza sativa* L. had a better salt-tolerant phenotype, and the hypocotyls were significantly longer than those of non-transgenic seeds after salt and drought treatment.

Index Terms— Chitosan nanoparticle;Drought Stress;*Eruca sativa*;*Oryza sativa*;Salinity Stress;*WRKY11* gene.

2 INTRODUCTION

Rice (*Oryza sativa* L.) is a cereal crop following Poaceae Family (grasses). Rice crop provides about half of the world population demands and considered the main source of energy. In areas such as Asia, Africa and Latin America where the demand for rice is a top priority, the population is expected to increase 1.5-fold by 2025. Egypt is considered one of the few countries to produce many varieties of rice cultivar. Rice is a crop grass which is grown more easily in the tropical regions ormay be conceded as semi-aquatic plant. Although it can grow in different environments, it grows faster and more vigorously in wet and warm conditions. This plant morphology consists of main stem and many tillers and may range from 0.6 to 6m (floating rice) in height. The tiller bears a ramified panicle that measures between 20 and 30 cm wide. Each panicle has 50–300 flowers (floret or spikelet), which form the grains. The fruit is a caryopsis type (Weising*et al.,* 1995; Sasaki, 2000; Gnanamanickam, 2009).In Egypt, the average of rice national yield is 9.4 tons/ha which is one of the highest all over the world. These high yields were achieved by spread of new short duration high yielding varieties (i.e., Giza177, Giza178, Sakha101, Sakha102, Sakha103, Sakha104, Giza182 and Egyptian Yasmine). Regardless of the limited land and water resources, increases in production and productivity have been accomplished through a well-coordinated interdisciplinary team. There is a significant reduction in the gap between yield potential and the national average (Badawi, 2004).

Most of Egyptian rice varieties produced higher grain yield with high watered-flooded soil. There is a clear constraint on reducing in Egypt based on the necessity to control salinity in the Northern Delta. The saline aquifer which underlines the Northern Delta will cause soil salinity problems due to upward migration when the hydraulic pressure gradient permits. Eng. Abdel Latif Khaled, head of the irrigation sector, at the Ministry of Water Resources and Irrigation, confirmed that the rice cultivated areas for a certain year were determined at the level of the Republic. The government estimated at 724 thousand and 200 acres in the governorates of Lower Egypt, and the ban on its cultivation in Upper Egypt, in cooperation with the Ministry of Agriculture. They stated that cultivating rice in non-authorized in certain areas due to the low irrigation and drainage. So, no cultivation of rice areas will be allowed outside the specified areas and grooves, and continuous campaigns will be launched to remove the seedlings of rice planted in the offending lands(Badawi and Ghanem, 1999; MENA, 2019).

Salinity and droughtare two major abiotic stress factors.Plant growth,development, and crop yield have greatly affected by high salinity and drought stresses. Recently, the physiological responses of plants to saline soils and drought conditions is essential.Genetic engineering tools are providing techniques to solve these problems. These stresses have affected osmotic and ionic stresses of plants and the physiological adaptations of plant growth parameters. When plants are exposed to drought and salinity stress, a series of genesare activated or inhibited, and the products of these genes may either further control the expressionof downstream genes or directly protect plants from stress damage. Drought reduction, via development of drought-resistant cultivars with higheryields suitable for water-limiting environments, enable to improve stablerice production. The development of drought-tolerant and high-yielding varietiesrequires a good knowledge of the physiological mechanisms, yield under droughtrelatedcomponents and the genetic control of traits contributing to drought resistance (Caplan*et al.*, 1990; Kumar*et al.*, 2016; Tang*et al.*, 2019).

Geneticengineeringtechnology in crop plantshasenhancedthe yieldproductivityto overcomeclimatechangebyintroducingdesirablegenetictraitstoagriculturalcrops.Effectivegenetictransformationi nplantsstillachallengebecause ofthecellwall as a barrier.Traditionaldeliverytechniquesareinefficient,damagingtotissue,orareonly effective inalimitednumberofplantspecies.Nanoparticlesarepromisingtechniqueforgenetic materialsdelivery,duetotheirabilitytopass throughplantcellwallswithoutexternalforceor disruptions. Chitosan is a polymer which has been used in nucleic acid delivery and tissue engineering applications. Chitosan is a natural linear cationic polysaccharide (Kumar *et al.*, 2004; Cunningham*et al.*, 2018).

This work introduces the WRKY11 gene to be transformed into *Oryza sativa* cultivar (Giza 177), in order to examine the tolerance against salinity and drought stresses. Gene transformation process was carried out via chitosan nanoparticles.

3 MATERIALS AND METHODS

Materials. Eruca sativa L. was the donor source of WRKY gene.Oryza sativa L. (Giza 177) (provided from Agriculture Research Center) was the receptor plant to be transformed with WRKY gene to resist abiotic stresses.

Methods

Obtaining of WRKY11Genes. The total genomic DNA of Eruca sativa was extracted using CTAB protocol described by Rogers and Bendich (1985). For PCR reaction 50ng of template DNA was used for each 25µl reaction. Each reaction mixture contained 12.5µl of 2X master mix (Biolene), 1µl of each forward and reverse primer (50nmole/base) (Biosearch) and complete up to 25µl by sterile dd H_2O. The WRKY11 primers were designed using snap gene® (2.3.3) software. These primers were WRKY11-F (TCTCCATTGTGTGTGAATGA) and WRKY11-R (GGCCACCAAGTCATTTATAT).

PCR reaction for this gene was performed for 40 cycles as follow: 30 min at 94°c, 30 min at 50°c and 1 min at 72°c. PCR product for the isolated WRKY11 gene was run on 1.2% (w/v) agarose gel comparing to (New England Biolab, #N3232S) ladder. The amplified PCR products were purified using GeneJET™ PCR Purification Kit (Thermo K0701).

Preparation of modified plasmid with WRKY11 Gene. The purified WRKY11 product was ligated separately to pMiniT Vector (NEB® PCR Cloning Kit, #E1202S. The kit was supplied with NEB 10-beta Competent E. coli (NEB #C3019)) following the manual instruction to obtain a modified pMiniT with WRKY11 gene.

Confirmation of Bacterial Transformation (Colony PCR). This was used to distinguish between recombinant and non-recombinant colonies; and performed using colony PCR. Bacteria with modified plasmid were grown on nutrient agar plates with 100 µg/L ampicillin. Separate colonies obtained from bacterial transformation were used as a template for PCR reaction. The primers, protocol and conditions for this reaction were as mentioned previously during WRKY11 gene detection from Eruca sativa. After that apply the product on 1.2% gel electrophoresis.

Sequencing of WRKY11 Gene. Samples from all clones were sequenced in GATC Company for sequencing which uses the ABI 3730xl DNA sequencer by using forward and reverse sequences, by combining the traditional Sanger with the new 454 technology. A nucleotide blast search was performed with the sequences using sequence alignment by NCBI blast (5) and Geneious software. The similarity/identity value determined from the pairwise comparison of the resulting genes sequence with the data on NCBI. Geneious compares the entered sequence to vast database of sequences and gives a percent value of similarity making it possible to confirm the presence of a given sequence.

Chitosan nanoparticle transformation. In this study, a fine powdered-mid molecular weight chitosan nanoparticle was used for transformation of the WRKY11 gene into the rice cultivar (Giza-177).

Preparation of chitosan-DNA nanoparticles (CS/pDNA). Mansouri et al. (2006) described the method of CS/DNA formation as follow: chitosan (CS) was dissolved in 25mM acetic acid, which was then adjusted to pH 5.5 at a final concentration of 0.08%. The CS and the pMiniT were first incubated in water bath at 55°C for 15min, respectively. Then CS–DNA complexes were prepared by adding the 0.08% CS to an equal volume of a pDNA solution (50mg/ml) followed by immediately intense stirring on a vortex mixer for 1min.

DNA retention and release. The binding of pDNA with CS was determined by agarose gel electrophoresis method (Li et al., 2003). CS/pDNA were loaded onto 1% agarose gel in Tris–borate/EDTA (TBE) buffer and electrophoresed at 60V for 40min. Then, gels stained with ethidium bromide were visualized with a UV/white gel documentation system.

DNA release study was carried out by Huang et al. (2005). Briefly, CS/pDNA NPs were centrifuged at 4°C, 14,000g for 30min, after which the pellet was re-suspended in equal volume of phosphate-buffered saline (PBS, pH 7.4) and incubated at 37°C with 100rpm agitation for 3 days.

Transfection of chitosan into plant tissue. The method of transfection of chitosan nanoparticle into plant tissue was developed by (Tawfik et al., 2018): germinated seedlings of rice were inoculated with 300µl of CS/pDNA mix. After that the explants were transferred on to MS medium containing (2mg/l BA and 1mg/l kin) hormones and 50µg/l kanamycin then incubated at 25±1°C for 4 weeks to regenerate plants.

Gene analysis

PCR analysis to Detect Transgenes. Total genomic DNAs were extracted from the leaves of transformed and untransformed rice lines using cetyl-trimethyl ammonium bromide (CTAB) method (Rogers and Bendich 1985). DNA fragments of WRKY11 transgenes were amplified using the total genomic DNA as templates by PCR.

Drought and Salinity Resistance Bioassay. Both transgenic and non-transgenic seedlings were transferred into peat-moss-soil and irrigated with 12% polyethylene glycol (PEG, for drought stress) and 0.02 and 0.06% sodium bicarbonates (for salinity stress); comparing to control (irrigated with distilled water).

Statistical analysis. The mean averages of datawere calculated using Minitab 19 software.

3. Results

Plant Material and growth conditions. Shoots of *Oryza sativa* (rice cultivar Giza 177) was regenerated on free MS media. The regeneration of these plants continued for three months to keep them until transforming the *WKR11* gene carried out.

Genes isolation,Cloning, Transformation into *E. coli*and Sequencing. DNA of *Eruca sativa* leaves were used as a source for isolation of *WKR11* gene using PCR (Fig. 1). The plasmid used for cloning in this work was pMiniT vector from NEB® PCR Cloning kit. Ligation products (plasmid backbone + WKR11 insert) were transformed into chemically 10-beta cells which illustrated in figure (2). The cells were propagated on a surface of LB medium supplemented with 50µg/L ampicillin to select the transformed clones.

Xu *et al*., (2014) mentioned that the *WRKY* family regulates the expression of genes closely related to both biotic and abiotic stresses. The functions of *WRKY* genes have been estimated in many plants, including *Arabidopsis*, barley, soybean, grapevine, sorghum, rice and maize.Hanna-Roseand Hansen (1996) proved that *WRKY11* gene expression improves rice resistance to salt stress. Also, reported that overexpression of DnWRKY11 improved the germination rate, root length, and fresh weight of transgenic tobacco under salt and drought stress when compared with the wild type. Also, Ali *et al*. (2018) transform both WRKY11 and WRKY17 genes into

Arabidopsis plant to be resistant to ABA, salt, and osmotic stress.

The sequenced gene was aligned to National Center for Biotechnology Information (NCBI) using BLAST tool. The alignment results showed that the *WRKY* sequence from *Eruca sativa* is new and not submitted yet.

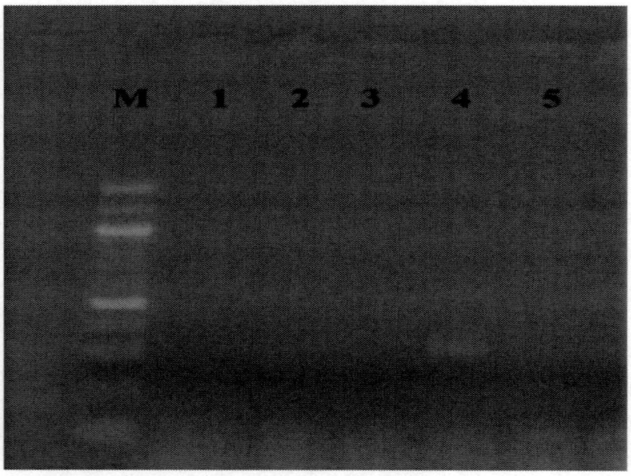

Figure 1. Gel electrophoresis for WKR11 gene from some Brassicaceae species

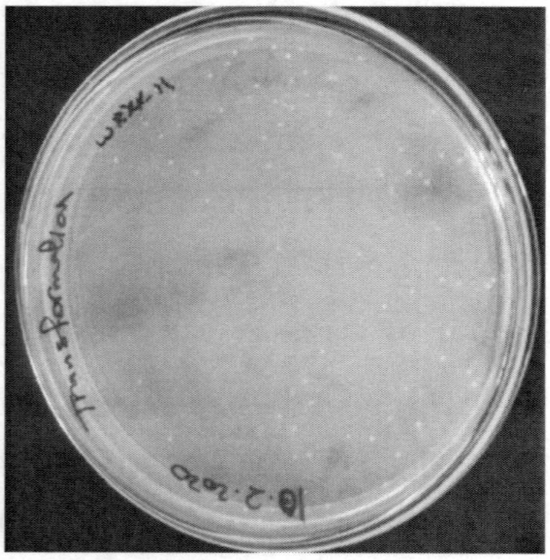

Figure 2. The transformed gene on LB medium with ampicillin (50µg/ml)

Confirmation of Bacterial Transformation (Colony PCR). In order to find correct ligation products, colony PCR was carried out by using colonies grown on LB medium supplemented with 50µg/L Kanamycin. PCR for a single colony for each gene was performed and the products were run on 1.2% agarose gel to show that both genes were cloned and transformed successfully. The results obtained were illustrated in figure (3) which showed bands and this indication for recombinant colonies.

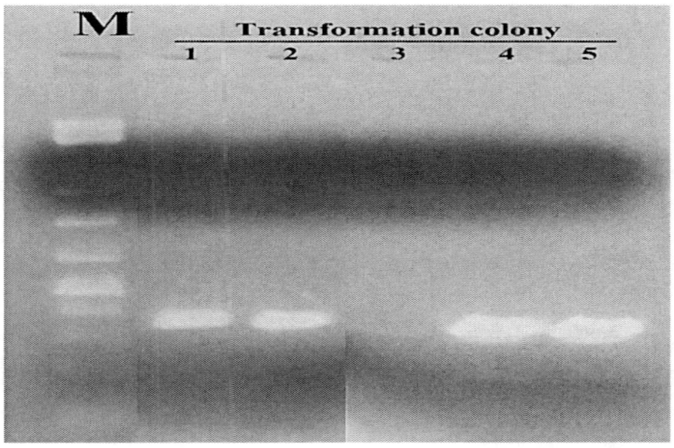

Figure 3.Gel electrophoresis of colony PCR product for WKR11 gene

Chitosan nanoparticles

Retention and Release. Chitosan (CS) polymers showed great ability to form complexes with DNA. The optimal ratio of complex formation between chitosan and pDNA was 1:1 and it was confirmed by electrophoresis of complexes on an agarose gel. The CS–DNA complex encapsulation capacity often depends on the molecular weight of CS. The results show that the extent of DNA retardation increases with an increasing ratio of CS/DNA (w/w) (Fig. 4). When plasmid DNA (pDNA) is mixed with chitosan, the electrostatic interaction drives the formation of complexes. Migration of pDNA on an agarose gel is retarded because of the charge neutralization and/or increase in molecular size of the complexes. The release of pDNA from CS-NPs was performed and examined in agarose gel electrophoresis as illustrated in figure (4). This release process is considered as simulation for the release of plasmid from chitosan nanoparticles due to the action of cell sap.

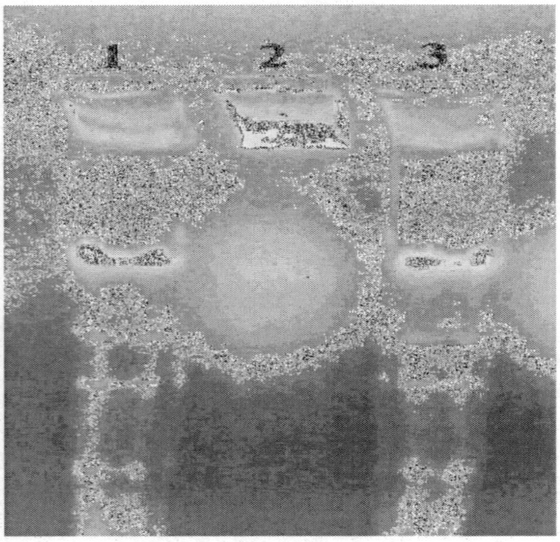

Figure 4. Gel electrophoresis of retention and release of plasmid from chitosan nanoparticles. 1: control free plasmid, 2: CS/pDNA complex, 3: plasmid released from CS nanoparticles

Transformation of WRKY11 Gene into Rice. In this experiment, CS was investigated as potential delivery systems for pDNA*in vitro*. The seedling explants of rice cultivar were immersed in the CS/pDNA for 30 min then transferred into MS media with 50µg/ml ampicillin. After four weeks incubation at 25°C under light conditions, the shoots are fully grown and multiplied. Then part of them was tested for gene transfer. The transformed rice lines with

WRKY11 genes via CS were shown in figure (5). These results agreed with Abdel-Razik*et al.* (2017) in using chitosan nanoparticles as delivery to thionin genes into potato plant tissues.

Figure 5. The transformed and non-transformed rice cultivar lines with the *WRKY11* gene via CS in tissue culture jars

Confirmation of gene transformation. The ampicillin-resistant clones of modified plasmid were subjected to PCR analysis to confirm the integration of the transgenes *WRKY11* into rice tissues. The DNA gel electrophoresis of the transformed rice cultivar was illustrated in figure (6). No amplified band was shown in PCR products from total DNA of non-transformed control rice cultivars as a negative control for transformation.

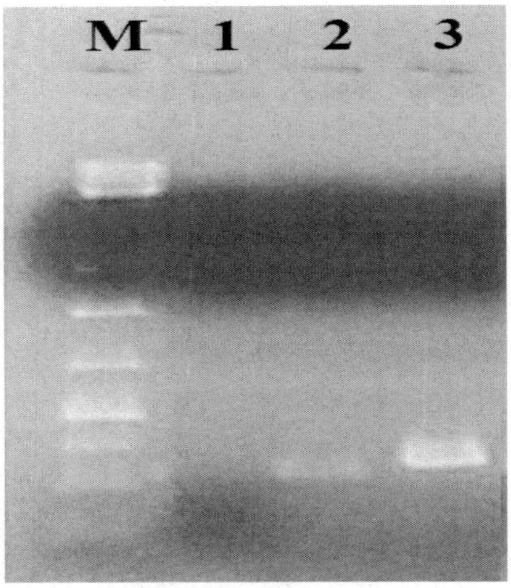

Figure 6. Gel electrophoresis for WRKY11 gene; M: marker, 1: non-transgenic rice, 2: transgenic rice and 3: control gene from *Eruca sativa*

Drought and Salinity Bioassay. In this bioassay, both transgenic and non-transgenic rice lines were exposed drought by irrigation with 12% PEG; and salinity by irrigation with 0.02M sodium bicarbonates. The total length of these cultivar lines was measured and indicated that the response of transgenic rice lines to growth under salinity and drought is obviously better than growth of non-transgenic ones. All that were indicated in figure (7) and table (1).

There is evidence that rice cultivation has actually improved some of the relatively saline soils in the Northern Delta. So that rice cultivation is very important conservation of soil fertility and reducing salinity hazard in particular areas such as Northern Delta (Badawi and Ghanem, 1999).Caplan*et al.* (1990) have isolated cDNA clones for a gene expressed primarily in root and sheath tissues of salt- or drought-stressed rice plants, and for a drought-induced peroxidase. Studies of these genes show that the molecular responses to adverse conditions can be tissue-specific, since neither gene was induced throughout the plant by either stress treatment.

Table 1. The mean average of total length "cm" of transgenic and non-transgenic rice lines in response to drought and salinity compared to control (dH_2O)

Transgene	dH_2O	PEG	0.02 M $NaHCO_3$
Non-transgenic	7.3	3.5	5.12
Transgenic	10.55	10.42	7

Figure 7. Resistance of transgenic and non-transgenic rice lines to drought and salinity comparing to control irrigated with water

Conclusion

In conclusion, EsWRKY11 was identified and characterized as a stress-inducible TF that responds to alkalinity, cold, drought, ABA and salt stress. EsWRKY11 improved the salt tolerance and Drought in *Oryza sativa* L. (Gize 177), which further confirmed the role of the WRKY gene family the salt stress response and its potential for engineering. Oryza sativa L. lines resistant to saline soil and by the use of WRKY genes we can reduce the weed plants to get more healthy rice crops as well as avoid the abiotic stresses, which helps us to sufficient our needs of rice and have the ability to export them out.

References

Abdel-Razik A., Hammad I., Tawfik E. (2017): Transformation of Thionin genes using chitosan nanoparticle into potato plant to be resistant to fungal infection. IOSR Journal of Biotechnology and Biochemistry, 3(3): 1-13.

Ali M.A., Azeem F., Nawaz M.A., Acet T., Abbas A., Imran Q.M., Shah K.H., Rehman H.M., Chung G., Yang S.H., Bohlmann H. (2018): Transcription factors WRKY11 and WRKY17 are involved in abiotic stress responses in *Arabidopsis*. Journal of Plant Physiology,226: 12-21.

Badawi T.A. (2004): Rice-based Production Systems for Food Security and Poverty Alleviation in the Near East and North Africa:New Challenges and Technological Opportunities.

Badawi T.A., Ghanem S.A. (1999): Water use efficience in rice culture. In: ChataignerJ. (ed.). Future of water management for rice in Mediterranean climate areas: Proceedings of the Workshops. Mon tpellier: CIHEAM. p. 39 -45 (Cahiers Option s Méditerranéenn es; n. 40)

Caplan A., Claes B., Dekeyser R., Van Montagu M. (1990): Salinity and Drought Stress in Rice. In: Sangwan R.S., Sangwan-Norreel B.S. (eds) The Impact of Biotechnology on Agriculture. Current Plant Science and Biotechnology in Agriculture, vol 8. Springer, Dordrecht. https://doi.org/10.1007/978-94-009-0587-0_24

Cunningham F.J., Goh N.S., Demirer G.S., Matos J.L., Landry M.P. (2018): Nanoparticle-Mediated Delivery towards Advancing Plant Genetic Engineering. Trends in Biotechnology, 1-16. https://doi.org/10.1016/j.tibtech.2018.03.009

Gnanamanickam S.S.(2009): Biological control of rice disease. Springer, Dordrecht Heidelberg London New York .pp1-11.

Hanna-Rose W., Hansen U. (1996): Active repression mechanisms of eukaryotic transcription repressors. Trends Genet, 12: 229–234.

Huang M., Fong C.W., Khor E., Lim L.Y. (2005): Transfection efficiency of chitosan vectors: Effect of polymer molecular weight and degree of deacetylation. Journal of Controlled Release, 106(3): 391-406.

Kumar A., Basu S., Ramegowda V., Pereira A. (2016): Mechanisms of drought tolerance in rice. Book chapter. Burleigh Dodds Science Publishing Limited.

Kumar M.N., Muzzarelli, R.A., Muzzarelli C., Sashiwa H., Domb A.J.(2004): Chitosan chemistry and pharmaceutical perspectives. Chem, 104: 6017–6084.

Li X.W., Lee D.K.L., Chan A.S.C., Alpar H.O. (2003): Sustained expression in mammalian cells with DNA complexed with chitosan nanoparticles. Biochimica et Biophysica Acta (BBA) - Gene Structure and Expression, 1630(1): 7-18.

Mansouri S., Cuie Y., Winnik F., Shi Q., Lavigne P., Benderdour M., Beaumont E., Fernandes J.C. (2006): Characterization of folate-chitosan-DNA nanoparticles for gene therapy. Biomaterials, 27(9): 2060-2065.

MENA (2019): Middle East News Agency. https://www.masrawy.com/news/news_egypt/details/2019/3/14/1531976

Rogers S.O., Bendich A.J. (1985): Extraction of DNA from milligram amounts of fresh, herbariurp

and mummified plant tissues. Plant Mol Biol, 5:69-76.

Sasaki T. (2002): Rice genomics to understand rice plant as an assembly of genetic codes. Current Sci. (83): 834- 839.

Tang Y., Bao X., Zhi Y., Wu Q., Guo Y., Yin X., Zeng L., Li J., Zhang J., He W., Liu W., Wang Q., Jia C., Li Z., Liu K. (2019): Overexpression of a MYB Family Gene, OsMYB6, Increases Drought and Salinity Stress Tolerance in Transgenic Rice. Front. Plant Sci. 10:168. doi: 10.3389/fpls.2019.00168

Tawfik E. (2018): Production of transgenic potato expressing antimicrobial thionin gene. *LAMBERT Academic Publishing. OmniScriptum Publishing Group, Latvia.*

Weising K., Atkinson G., Gardner C. (1995): Genomic fingerprinting by microsatellite-primed PCR: a critical evaluation. PCR Methods and Applications, 4: 249-255.

Xu X.B., Pan Y.Y., Wang C.L., Ying Q.C., Song H.M., Wang H.Z., et al.(2014): Overexpression of *DnWRKY11*enhanced salt and drought stress tolerance of transgenic tobacco. Biologia, 69: 994-1000.

Made in United States
North Haven, CT
04 November 2026

21731729R00017